Josiane Elias Nicolodi
Roberto Nicolodi

The concept of division, reflections on learning

Josiane Elias Nicolodi
Roberto Nicolodi

The concept of division, reflections on learning

From the perspective of students in the second year of primary school

ScienciaScripts

Imprint
Any brand names and product names mentioned in this book are subject to trademark, brand or patent protection and are trademarks or registered trademarks of their respective holders. The use of brand names, product names, common names, trade names, product descriptions etc. even without a particular marking in this work is in no way to be construed to mean that such names may be regarded as unrestricted in respect of trademark and brand protection legislation and could thus be used by anyone.

Cover image: www.ingimage.com

This book is a translation from the original published under ISBN 978-3-330-99676-2.

Publisher:
Sciencia Scripts
is a trademark of
Dodo Books Indian Ocean Ltd. and OmniScriptum S.R.L publishing group

120 High Road, East Finchley, London, N2 9ED, United Kingdom
Str. Armeneasca 28/1, office 1, Chisinau MD-2012, Republic of Moldova, Europe
Managing Directors: Ieva Konstantinova, Victoria Ursu
info@omniscriptum.com

Printed at: see last page
ISBN: 978-620-8-60135-5

I dedicate this book to my husband, Roberto, my daughters, Júlia Beatriz and Rayane, and my parents, Onécio and Maria.

SUMMARY

The aim of this research was to characterise the written solutions used by primary school students when solving division problems. Based on the theoretical-methodological frameworks proposed by Vergnaud and Nunes, Campos, Magina and Bryant, an investigation was undertaken to answer the following research question: What understanding of division is revealed in the written solution of primary school students when solving division problems? Data collection was carried out with 38 children from a public school in Navegantes, where they were asked to solve a form with division problems, exact and inexact, partitioning and quotas, in order to obtain information about the object being investigated. The qualitative analysis of the subjects' written solutions to the division problems resulted in a map that lists the stages that led to the strategies that resulted in the schemes used by the children to solve the division problems. And also in the coordination of the dividend, divisor and quotient factors involved in the situation arising from the problem statement. It can be concluded that the understanding presented in the expected solution to the problems is not directly linked to the classification of the problems, but rather to the situation described in the statement.

Keywords: Schemes; Division; Solutions.

SUMMARY

CHAPTER 1

INTRODUCTION

In order to situate this work and justify the choice of topic, we will present the current situation of maths teaching as revealed by the Ministry of Education (MEC) and international organisations. In recent years, the Brazilian organisations responsible for education have been concerned with reversing the situation of education in Brazil.

In 2004, Mônica Weinberg published the article "Assim vai Mal" in Veja magazine, in which she commented that Brazil was among the worst in the education ranking, specifically in mathematics, with the last place, data revealed by the survey carried out in 2003 by the Programme for International Student Assessment (PISA), organised by the OECD (Organisation for Economic Co-operation and Development), in which students from fifty-seven countries took part. The data is collected every three years to assess the level of students aged 15 on average in three basic skills: reading, maths and science, with a focus on problem-solving. The same assessment is applied in all countries and it was observed, on a scale of six levels, that half of the young Brazilians interviewed scored below grade 1. In maths, 53% of the questions were wrong. In the survey carried out in 2003, Brazil came last, revealing that young people are leaving primary school without knowing how to add and subtract, lagging behind young people from poorer countries such as Tunisia and Indonesia. The results of the 2006 PISA, published by INEP (The National Institute for Educational Studies and Research Anísio Teixeira, 2006) show similar data to the 2003 results.

Concerned about these alarming figures, in 2005 the Ministry of Education introduced the Prova Brasil, which assessed knowledge of Portuguese language and maths (also focusing on problem solving). Students from grades 4 and 8 in public schools took part· The results presented by INEP (Instituto Nacional de Estudos e Pesquisas Educacionais Anísio Teixeira), which organised the tests, were alarming. The data reveals that, on average, Brazilian education is far from a minimum standard of quality. The IDEP (Basic Education Development Index) index for Brazil in 2005 was 3.80, in 2007 it was 4.3 and the target for reaching 6.0 is only for 2021.

Education in the early grades of primary school is a privileged space for literacy, and in particular for mathematical literacy, which requires the construction of elementary mathematical concepts, such as those presented in the PCNs (Brazil/MEC1997) de Matemática: natural numbers and decimal numbering system, operations with natural numbers, space and shape, magnitudes and measures, information processing. This requires the use of teaching strategies that facilitate and organise children's written solutions, such as problem solving.

Extremely important for learning mathematical concepts, problem solving makes it possible to present mathematical concepts in situations that are familiar to the student, in other words, it brings theory and practice closer together so that the concepts have meaning for children. According to Piaget & Szenúnska (1971), children build their mathematical foundations through the need to solve problems in their own time, imposed in everyday situations. Therefore, like "primitive" man, it starts from a sense of number to abstract construction, a construction in which the time factor occupies an important place. For human beings to develop logical mathematical thinking, they need to make all possible relationships between objects: they are the same, they are different, they are bigger, they are smaller and so on. Thus, Piaget (1970, p. 13) argues:

> Verbal or reflected intelligence is based on practical or sensory-motor intelligence, which in turn draws on acquired habits and associations in order to recombine them. On the other hand, these same habits and associations presuppose the existence of the reflex system, whose connection with the anatomical and morphological structure of the organism is evident.

Thus, for the author, intelligence can be defined as one of the manifestations of life, in other words, a form of adaptation. Thus, at all levels of development, cognitive behaviour is an action (concrete or internalised), the function of which is to adapt the subject to their environment, through interaction with it. This development is continuous, as we have, on the one hand, the notion of action and, on the other, that of function: through processes of assimilation and accommodation, the subject gradually coordinates their actions at an increasingly higher level of structural complexity, according to Piaget (1970). This is no different in maths learning.

From a pedagogical point of view, it is extremely important for the teacher to get the child to build all the possible relationships between objects, in the constructions of their own play: grouping objects by their similarity; making simple and serial classifications; comparing sizes: bigger, smaller, equal and others, will allow the construction of mathematical knowledge as described in the PCN's (Brazil/MEC1997).

According to Piaget (1996, p. 18), "most schemes, instead of corresponding to a finished hereditary assembly, are built up little by little, and even give rise to differentiations, by accommodation to modified situations or by combinations". The meaning that Piaget gives to the notion of schema, or action schemas, is that which is common to various applications or repetitions of the same action, in other words, we realise that the definition of schemas, and this definition becomes a fundamental framework for other theories that derive from Piagetian theory, such as Vergnaud's theory, on which we will base this work.

With regard to the construction of mathematical concepts, Piaget (1996) describes that logical-mathematical construction is neither invention nor discovery; it is a process of reflexive abstractions and the construction of new combinations.

Brun (1996, p. 18) describes in his work that, because Piaget was concerned with the epistemology of maths, he ended up directing his writings towards the pedagogical field, and thus added to epistemology considerations and recommendations aimed solely at teaching.

Thus, it is clear that by building spontaneous knowledge alone, children cannot fulfil the needs imposed by society throughout their lives. Therefore, the role of the school is to help them transform this spontaneous knowledge into scientific knowledge, linking these two forms of knowledge. As such, we educators interfere in this process of building our students' knowledge. For this reason, the act of teaching involves a broader understanding than the teacher's restricted space in the classroom, or the activities carried out by the students: we need to identify the knowledge already established by our students, in their development, in the various experiences of their lives.

Therefore, we start from the assumption that, before entering school, children have spontaneous knowledge of various mathematical contents, and among these contents is the object of study of this research, the mathematical concept of division (Lautert and Spinillo, 2002). We chose this object because we have seen that division is one of the most difficult obstacles for children when learning maths in the early primary school years and, at the same time, it is crucial for the construction of concepts that are learnt later, such as fractions.

Vergnaud (1991) considers division to be one of the most complex of the four operations, for various conceptual reasons: it is not always exact, the quotient is not always the result of applying the operator to the operand, there can be non-zero remainders, division as an operative rule is not always the inverse of multiplication. Division is also related to two different ideas, dividing and measuring, with the first, partitioning, being more emphasised than the second, by quota.

Research from a constructivist perspective has shown that children recognise the complexity of the mathematical concept of division, find it easier to work with partition problems, have difficulties dealing with the remainder and younger children find it easier to use paper and pencil strategies or mental calculations to work with the remainder than to use materials (Selva 1998). Other studies have described the initial construction of the concept of division by the child, where it is important for the child to recognise these two classes of problems: division by partitioning and division by quotas (Corrêa, 2004), and have noted how early division is among children (Moro, 2004).

Based on the above assumptions, it is assumed that the teacher needs to know the children's solutions when solving problems involving division, because the strategies used by the children are based on the notions they have built up about this mathematical concept in their daily experiences. Therefore, it is necessary to know what knowledge children already have about division before they are taught this content at school, what strategies they use to solve division problems and what notations they produce to record their solutions.

So the question of this research is: *What understanding of division is revealed in the written solution of primary school students when solving division problems?*

In order to respond to the problem presented, the general objective of this research is to **characterise the written solutions used by primary school students when solving division problems.** In order to meet this objective, the fundamental aspects of these solutions were content, strategies and schemes, which allowed the following specific objectives to be formulated:

1. Check that the written solutions contain the terms of the division (dividend, divisor) and the result (quotient), and that they are consistent with the information in the problem statement;

2. Identify whether students in the first grade of primary school use the "equitable distribution" and "one-to-many correspondence" action schemes (according to Nunes, Campos, Magina and Bryant, 2005) in their written solutions; and

3. Check whether the written solutions show coordination between the schemes.

The researcher hopes that this research can suggest ways for teachers to analyse students' written solutions to problems involving the concept of division, seeking to understand their strategies and, if possible, infer the schemes that led students to different solutions.

To help understand, discuss and analyse the proposed topic, the following topics will be reviewed in the literature: Conceptual Fields Theory, Multiplicative Structures, Division and Schemes.

CHAPTER 2

CONCEPTUAL FIELDS

Gérard Vergnaud (1998, p. 23), a French psychologist, scholar and disciple of Piaget, seeks to broaden and direct the Piagetian theory of logical operations, of the structure of thought, towards reflection on the cognitive functioning of the "subject in situation". He centred his studies on the knowledge in question and the conceptual analysis of the domain of that knowledge, unlike Piaget who focused his research on the structuring of thought. As a milestone in Piagetian theory, the concept of schema also became fundamental in Vergnaud's theory.

Vergnaud (1990) improved the definitions of concept, because when we become interested in learning and teaching this concept, we realise that a single concept does not take on meaning in a single situation and that this situation cannot be analysed through a single concept. For the author (1990), the operationality of a concept needs to be proven through various situations, for example, the concept of function is only understood through various practical problems, which make it possible to apply the properties according to the situations, understood in the course of its learning.

Based on these principles, Vergnaud (1990) developed the theory of conceptual fields, about which Brun (1996, p. 22) states "that it most closely resembles the psychology of concepts":

> "... an organisation of knowledge which does not allow itself to be immediately enclosed in descriptions of knowledge and which also takes into account the ongoing activities of the cognising subject in a situation. Let's not forget that action in a situation is the source of concept formation." (BRUN, 1996, p. 22)

Vergnaud (1996b) developed the theory of conceptual fields in order to better understand specific development problems within the same field of knowledge. To do this, he was inspired by his predecessors, above all Piaget and Vygotsky, both of whom were interested in a theory of conceptualisation. Vygotsky was mainly interested in the role of language and symbolic forms and Piaget always focussed his studies on logical structures and the development of thought operations, according to Vergnaud (1998).

According to Vergnaud (1990), knowledge is organised into conceptual fields and in order for children to master these organisations, they need the passage of time so that, through their experiences, maturity and level of learning, they can model their organisations. According to the author, a conceptual field is:

> "an informal and heterogeneous set of problems, situations, concepts, relationships,

structures, contents and thought operations, connected to each other and probably intertwined during the acquisition process. Mastery of a conceptual field does not occur in a given time, it can take a few months or even a few years" (VERGNAUD, 1990, p 136).

The author describes that he developed "the theory of conceptual fields to try to better understand the specific developmental problems within the same field of knowledge" (VERGNAUD, 1996b, p. 11). Developing the theory of conceptualisation. Although it was developed to explain the process of conceptualising additive and multiplicative structures, the theory of conceptual fields is not specific to mathematics, but also applies to other sciences.

Vergnaud (1990) cites the example of the conceptual field of multiplicative structures, which consists of all the situations that require multiplication, division or a combination of these operations. In this way, various mathematical concepts are involved in the situations that make up the conceptual field of multiplicative structures and in the thinking required to master these situations. These concepts include linear functions, fractions, ratios, rates, rational numbers, multiplication and division, according to Vergnaud (1990). The same applies to the conceptual field of additive structures "which is the set of situations whose mastery requires addition, subtraction or a combination of these operations" (VERGNAUD, 1990, p. 146).

The operations of additive structures are made up of systems of various types of schemes, applicable to different situations, which are progressively coordinated in a complex construction at different psychogenetic levels, according to Vergnaud (1990).

For Vergnaud, knowledge is organised into conceptual fields, where mastery is the responsibility of the subject and takes place over a long period of time. "It is a psychological theory of the process of conceptualising reality that allows us to locate and study continuities and ruptures between knowledge from the point of view of its conceptual content" (1996b, p.ll).

> the theory of conceptual fields is a complex theory, because it involves the complexity arising from the need to encompass in a single theoretical perspective the entire development of progressively mastered situations, the concepts and theorems needed to operate efficiently in these situations, and the words and symbols that can effectively represent these concepts and operations for students, depending on their cognitive levels (VERGNAUD 1994, p.43).

Among the relevant concepts of the theory of conceptual fields, we will pay more attention to the definition given by Vergnaud to the concept of conceptual field, the concept of schema (invariant organisation of behaviour for a given class of situations, in the sense used by genetic epistemology), inherited from Piaget, the situation, the invariants and his conception of concept, to support the possibility of communication in learning mathematics at school.

Vergnaud defines a conceptual field as "a set of situations whose mastery requires, in turn, the mastery of various concepts of a different nature, in other words, a study to make sense of the difficulties observed in conceptualising reality" (1990, p. 146). As already mentioned, various mathematical concepts are involved in the situations that define the conceptual field of multiplicative structures and the conceptual fields do not necessarily have to be independent. In order to understand one concept, it is often necessary to understand the other. As an example, we have the conceptual field of multiplicative structures, "where every situation that can be analysed through problems of simple and multiple proportions, which normally result from multiplication or division, or both operations, are part of the conceptual field of multiplicative structures" (1990, p. 146).

2.1 Concepts

Vergnaud defines concepts with three sets (S, I and R):

S is a set of situations that give meaning to the concept; I is a set of invariants (objects, properties and relationships) on which the operationality of the concept rests, or the set of operative invariants associated with the concept, or the set of invariants that can be recognised and used by subjects to analyse and master the situations in the first set; and R is a set of symbolic representations (natural language, graphs and diagrams, formal sentences, etc.) that can be used to indicate and represent these invariants and consequently represent the situations and procedures for dealing with them. (1990, p. 145)

With this, we realise that in order to study the development and use of a concept, throughout its learning or use, it is necessary to consider these three sets simultaneously. "In general, there is no two-way correspondence between signifiers and signifieds, nor between invariants and situations; meaning cannot therefore be reduced to either signifiers or situations" (VERGNAUD, 1990, p. 146).

Concepts become meaningful through situations, so we realise that it is situations and not concepts that are the main input to a conceptual field. "A conceptual field is, in the first place, a set of situations..." (VERGNAUD, 1990, p. 146).

2.2 Situations

The concept of situation used by Vergnaud (1990) is the meaning normally attributed by psychologists, that the subject's cognitive processes and responses are a function of the situations they face. The author defines two main ideas in relation to the meaning of situation: variety and history, so in a given conceptual field we have a variety of situations that gradually shape the students' knowledge. Thus, "many of our conceptions come from the first situations we were able to master or from our experiences trying to modify them" (1996b, p. 10).

However, for a concept to become meaningful, it has to go through various situations, and it is these situations that give the concept meaning, according to Vergnaud (1994). And this meaning is the relationship that the subject organises with situations and signifiers. Thus, "it is the schemes, behaviours and their organisation, evoked in the subject by a situation or a signifier (symbolic representation) that constitute the meaning of that situation or that signifier for that individual" (VERGNAUD, 1990, p. 158). Let's use the example of the meaning of division for a subject, which is the set of schemes they use to deal with situations they are faced with and which result in the idea of division. It may not be necessary for the subject to resort to all the schemes they already have, which they have acquired in situations familiar to them, or it may even be necessary to organise other schemes according to the situation. This is a subset of the schemes that the subject has, or the possible schemes according to Vergnaud (1990).

Let's see that the idea of a conceptual field is related to a triplet (referent, signified and signifier); however, as it is situations that give meaning to the concept, we arrive at the concept of situation and from there to that of schema. The concept of schema, as we shall see, will lead us to the concept of operant invariant.

We note that the first set of situations is related to the concept, the second to the operative invariants is what gives meaning to the concept and the third of symbolic representations is the signifier.

> This implies that in order to study the development and use of a concept, throughout its learning or use, it is necessary to consider these three sets simultaneously. In general, there is no two-way correspondence between signifiers and meanings, nor between invariants and situations; meaning cannot therefore be reduced to either signifiers or situations (VERGNAUD, 1990, p. 146).

According to Vergnaud, the conceptions we have today are the result of situations that we have managed to master or that have been attributed to us by our experiences, and we try to modify them. It is the situations that give meaning to the concept, and the meaning is not in the situations any more than it is in the words and symbols, according to Vergnaud (1990).

2.3 *Schemes*

For Vergnaud (1990), a schema is the invariant organisation of behaviour for a given class of situations. For him, it is in the schemas that we can see the subject's knowledge in action, in other words, the cognitive elements that make the subject's action operative, so several schemas can be seen in succession.

According to Vergnaud (1994), schema is the concept introduced by Piaget to account for the ways in which both sensory-motor skills and intellectual skills are organised. In organising a schema, we generate actions and have rules, which cannot be used in a standardised way, as the sequence of actions depends on the parameters of the situation, according to the author. A schema can be used efficiently for different situations and can generate different sequences of action, information gathering and control, depending on the characteristics of each particular situation.

Schemas can be more or less elaborate, for example the schemas used by children or adults, but this does not interfere with the central idea of schemas as being the structural form of the activity or the subject's invariant organisation of situations, as described by Vergnaud (1990). In this way, we can identify the action schemas that are present in the solutions to division problems written by students in the first grade of primary school, in order to highlight the schemas pertinent to division and facilitate the learning of this mathematical concept.

Let's go back to the definition of schema: "schema is the invariant organisation of behaviour for a given class of situations"(VERGNAUD, 1990, p. 136). To understand this definition, let's get to know the ingredients of schemas, according to Vergnaud (1990, p. 136):

> 1. anticipations of the goal to be achieved (in the schema there is always a class of situations in which the subject discovers the purpose of their activity, i.e. the expected effects);
>
> 2. action rules of the type if ... then they are the ones that constitute the schema, and allow the continuity of the subject's sequence of actions;
>
> 3. operative invariants (theorems-in-action and concepts-in-action) that direct the subject's recognition of the elements pertinent to the situation, classified as knowledge contained in the schemas;
>
> 4. possibilities of inference (or reasoning) that allow calculation, rules and anticipations based on the information and operative invariants available to the subject.

According to Vergnaud, when the subject uses an ineffective schema for a given situation and experiences lead them to change their schema or modify it (1990, p. 138). We return to the Piagetian idea that schemas are at the centre of the process of adapting cognitive structures, in assimilation and accommodation. However, Vergnaud gives the concept of schema a much broader scope than Piaget and insists that schemas must be related to the characteristics of the situations to which they apply.

An example of a schema, cited by Vergnaud (1996b, p. 6), is the schema of a five-year-old child's enumeration of a small collection of discrete objects. 6), is the scheme of enumeration of a small collection of discrete objects by a five-year-old child: however much the way of counting varies, for example, chairs in the living room, pens on the table, friends in the class, there is still an invariant organisation for the scheme to work: coordination of eye movements and finger and hand gestures, correct enunciation of the number series, identification of the last element in the series as the cardinal of the enumerated set (accentuation or repetition of the last "number" pronounced). It is easy to see that the scheme described makes use of perceptual-motor activities, signifiers (number words) and conceptual constructs such as the two-way correspondence between sets of objects and subsets of natural numbers, cardinal and ordinal and others. It also makes use of knowledge, such as that which identifies the last element of the ordinal series with the cardinal of the set.

As mentioned above, Vergnaud refers to the concepts contained in schemas as "concepts in action" and "theorems in action" and, in a broader definition, as "operative invariants". The author uses the following situation for 13-year-old students as an example of a theorem in action:

> "The average consumption of flour is 3.5kg per week for ten people. How much flour do fifty people need for 28 days? A student's answer: 5 times as many people, 4 times as many days, 20 times as much flour; therefore, 3.5 x 20 = 70kg".
> Vergnaud, (1994, p. 49)

According to Vergnaud (1994, p. 49), "it is impossible to realise this reasoning without assuming the following implicit theorem in the student's head: f(nlxl, n2x2) = nln2 f(xl, x2), i.e. consumption (5 x 10, 4 x 7), consumption (10, 7). For the author, this theorem is correct because the ratio 28 days to 7, is the same ratio as 50 to 10 people in this situation, so this theorem is easily applied in this situation, which would not be the case with other numerical values.

Therefore, by making it possible to treat the four operations of classical arithmetic as conceptual structures in a more consistent way, and by providing space to study their psychogenetic interrelationships, the theory of conceptual fields is of great interest to school teaching because it makes it possible to better analyse the dialectical relationship that occurs there between action, practical situations and theoretical verbalisation (Vergnaud, 1990).

CHAPTER 3

MULTIPLICATIVE STRUCTURES

According to Vergnaud, "the knowledge that children acquire must be constructed by themselves, in a relationship that they are able to make about reality, that they are able to perceive, compose and transform the concepts that they progressively construct" (1991, p. 9).

And mathematical concepts, according to Vergnaud, "form a set of notions, relationships, systems of relationships that support each other" (1991, p. 10), and the way in which the teacher presents this to the children is crucial to their learning.

According to the PCN (1998), the basic operations that need to be worked on gradually from the early grades of primary school are the four classic operations: addition, subtraction, multiplication and division. For this reason, various studies have been carried out to make maths education more efficient and meaningful. Much research has been carried out into the nature of the concepts and relationships that mark the nature of these operations, as well as the ways in which students understand them, as contributed by Vergnaud (1990).

Vergnaud centred his studies on additive and multiplicative structures in order to study the difficulties that students have in these areas. The conceptual field of multiplicative structures includes any situation involving multiplication, division or both operations simultaneously. In this way, we have mathematical concepts such as: proportion problems, linear and non-linear functions, fractions, ratios, vector space, dimensional analysis, rates, rational numbers, multiplication and division, which according to Vergnaud (1990), are contained in the situations that make up this conceptual field.

However, there are already several studies on the development of concepts and situations involved in solving multiplication and division problems. Two essential types of multiplicative relationships are considered: those that involve multiplication and those that involve division. It is from Vergnaud's (1990) perspective that the various concepts and relationships involved in the various forms of multiplication and division constitute a conceptual field, that of multiplicative structures.

According to this perspective, while most multiplicative school problems involve a quaternary relationship between measures, there are those that involve a ternary relationship between quantities, and these are called product of measures problems.

Vergnaud (1990) points to two forms of multiplicative relations: the isomorphism of measures and the product of measures (Cartesian product). According to the author, the isomorphism of measures is a quaternary relationship in the presentation of a problem, i.e. one in which two quantities are measures of a certain type and the rest are measures of a different type. At school, most problems have a quaternion relationship, which is often used to introduce or practise the concept of multiplication. Measurement isomorphism problems are commonly known by teachers as multiplication problems of the successive sums type.

So, in multiplication, the aim is to find a "product measure" by combining two or more given elementary measures. In division, the aim is to find an elementary measure from another combined with a product measure.

The importance of the multiplicative relationships present in product-of-measures problems lies in the fact that underlying the development of these relationships are schemas and systems of schemas that organise human cognition. In division, we have one-to-one and one-to-many correspondences and equitable distribution.

Among Piaget's many contributions to education, and especially to maths education, is the idea that the understanding of additive operations originates in children's schemes of action (Piaget & Szeminska, 1971). We note the progressive construction of these modes of reasoning in young children, with expressive logical-formal changes during adolescence. There is therefore not only the possibility but also the need to activate these constructions during the school years, in the most varied circumstances of solving arithmetic problems and interpreting reality.

The aim of this research is to characterise the forms of written solution used by primary school pupils when solving division problems, so it is an investigation into multiplicative structures.

Nunes, Campos, Magina and Bryant (2005) dedicate a chapter of their book to a conception of multiplicative structures in the classroom. According to these authors, the idea that is being transmitted in educational practice, that multiplication is a sum of equal parts, is no longer the only alternative for teaching the mathematical concept of multiplication. For them, the relationship between addition and multiplication is not conceptual. This relationship is due to the fact that the process of calculating multiplication can be done through addition, because multiplication is distributive in relation to addition.

"Additive reasoning refers to situations that can be analysed on the basis of a basic axiom : the whole is equal to the sum of the parts (NUNES et al., 2005, p. 84)". Additive reasoning is summed up in this statement, and around it we have various possibilities: if we need to know the total, we add up the parts; if we want to know the value of one of the parts, we subtract the other part from the whole; to compare two quantities, we analyse what is left over when we remove the other quantity from it. Therefore, the part-whole relationship is the conceptual invariant of additive reasoning.

"Any multiplicative situation involves two quantities in a constant relationship with each other." (NUNES et al., 2005, p. 85), so the invariant of multiplicative reasoning is the existence of a fixed relationship between the variables.

CHAPTER 4

THE DIVISION

Division is one of the four basic and fundamental mathematical operations (addition, subtraction, multiplication and division) and determines the number of times a number (divisor) is contained within another number (dividend). Division is the inverse operation of multiplication, just as subtraction is of addition, but this reciprocity does not make it easy for students to learn and understand. It can be said, through experience of teaching this concept, that the reason for this difficulty is the use of its algorithm, coupled with the lack of contextualisation of division situations - problem situations.

According to Vergnaud (1991), the operation of division involves complex operative rules, such as the use of successive divisions, multiplication, subtraction, or even the search for a quotient that can involve a remainder and result in fractional numbers. In addition, division requires students to establish different relationships, such as considering the size of the whole, the number of parts, the size of the parts, which must be the same, the direct relationship between the total number of elements and the size of the parts, and the inverse relationship between the size of the parts and the number of parts. This diversity can be contextualised through problem solving, one of the guiding principles of the PCN for Primary School Mathematics.

According to the PCN (1998), in relation to basic operations, the focus should be on understanding the different meanings of each one, the relationships between them and the study of calculation (exact and approximate, mental and written). In addition, the document mentions the importance of problem situations in understanding the existence of numbers and operations, as well as the study of issues that make up the history of the development of mathematical knowledge. One of the guiding principles of the PCN (1998, p. 57) states that "mathematical knowledge is historically constructed and is therefore constantly evolving". Maths teaching should enable students to "recognise the contributions it makes to understanding information and positioning themselves critically in the face of it".

The division operation involves knowledge beyond that related to obtaining equivalent parts when dividing. As a multiplicative operation, it requires coordinating the factors involved - dividend, divisor and quotient - by understanding the relationships that these terms can establish between themselves (CORRÊA, 2000, p 05).

Therefore, the

> Research into the relationship between children's everyday experience of sharing and their intuitive knowledge of division indicates that this experience, although necessary, is not sufficient for children to understand the relationships established between the terms involved in division situations (CORRÊA, 2002, p 04).

From an informal point of view, division is the act of dividing, apportioning, separating the parts of a whole:

- Dividend: the "whole" (total), which you want to distribute in equal parts;
- Divisor: delimits the number of parts to which the whole should be distributed;

The result and the rest:

- Quotient: the corresponding quantity in each of the parts that distributed the whole;
- Remainder: the quantity left over, i.e. not enough for another round of distribution.

Among the problems that relate to division are exact and inexact, partitioning and quotas.

- Exact division problems: when the remainder is equal to zero;
- Inexact division problems: when we have a non-zero remainder;
- Partition problems: we must distribute the whole into equal parts;

Quota problems: we need to establish the divisor.

The degree of difficulty of these problems varies, as is the case with the isomorphism problems known as division by partition and division by quotas, which we will focus on in this research. We will therefore use examples such as those used in the study by Lautert and Spinillio (2002) for these classifications of division problems.

"I paid £16 for four bracelets, what is the price of each bracelet?"
Lautert and Spinillio (2002)

"Júlia bought 15 candies and had five boxes. She wanted to put the same number of candies in all the boxes. How many candies did she have to put in each box?"
Research problems form (Appendix 2)

These division problems are classified as partition division problems, where we have the initial quantity and the number of times (number of parts) this quantity must be distributed in order to find the size of each part (number of elements). In these situations, for children to be able to solve these problems they need to establish the part-whole relationship, i.e. they need to know that the quotient to be obtained refers to the number of parts (price and candies), that the dividend is represented by the whole (R$16.00 and 15 candies), and that the divisor refers to the number of parts into which the whole will be divided (four bracelets and five boxes). In both problems, the fixed ratio is unknown, in the first the ratio to be discovered is four reais per bracelet and in the second three candies per box.

In division by quota problems, an initial quantity is given, which needs to be divided by pre-established quotas (the size of the parts). Examples:

"I have R$16.00 and I want to buy some bracelets that cost R$4.00 each.
How many bracelets can you buy with that amount of money?"
Spinillio and Lautert (2002)

"Júlia bought 15 candies and wanted to put five candies in each box. How many boxes will she need?"
Research problems form (Appendix 2)

In this case, for children to be able to solve these problems they need to consider that the quotient to be obtained refers to the number of parts into which the whole has been divided, that the dividend is represented by the whole and that the divisor refers to the size of the "quota" parts, i.e. the number of bracelets and boxes. Therefore, in this situation, the fixed ratio is known (R$4.00 per bracelet and five candies per box).

These problems mentioned above are very similar, as they have the same numerical representations, but they cannot be considered to be of the same nature, because if we change the unknown to be determined, we consequently change the nature of the operation to be applied when solving. With these examples, we can confirm that there are various situations that arise from mastering different properties in order to understand the same concept, one of the main statements that Vergnaud mentions in his theory.

We can see in the literature that partitioning problems are considered easier by children than division by quota (Selva, 1998). When children start school, they already have the notion of distributing quantities in equal parts until it is no longer possible to distribute them, a notion they have acquired through situations they have already experienced. Perhaps this is why division by partitioning is considered easier. Notions about division stem from the idea of distributing, as Selva shows.

According to Nunes, Campos, Magina and Bryant, (2005), in partition division problems we have the action of distributing equal quantities between the elements, based on one-to-one correspondence, until there are no more quantities to be distributed or the remaining quantity is insufficient to be distributed, while in quota division problems the resolution process begins based on the size of each part. These actions are classified as schemas in Vergnaud's theory.

4.1 The action schemes pertaining to the Division

In fact, the everyday sharing situations that young children face can be modelled, from a mathematical point of view, by the division algorithm. "In terms of the action schemes involved, these same situations can be related to the division operation through the use of term-by-term correspondence and the notion of equivalence," according to Correa (2000, p. 04).

Among the schemes used by children in solving division problems in the literature, as mentioned above, we highlight equitable distribution (notion of equivalence), which is the action of distributing equal quantities among the elements, and in a less elaborate action, we noticed one-to-one correspondence, that is, one for each one, until there is no more to distribute or the quantity to be distributed is insufficient for another round of distributions. And the one-to-many correspondence (term-by-term correspondence), which is done when the divisor is unknown, separating the whole into established quantities and thus finding the total number of parts.

When distributing, children mainly use correspondence schemes in order to establish equivalence between the parts. In this way, children can only use procedures involving addition where all they have to do, for example, is repeat the same set of actions until there are no more elements available for a second distribution. In this process, equivalence is achieved by adding or subtracting a few elements to be distributed (CORRÊA, 2000, p. 05).

According to Nunes (et al., 2005), children between the ages of 4 and 5 do not know how to coordinate the schemes in action that give rise to the concepts of multiplication and division. In their studies, the schemes observed were one-to-many correspondence and equitable distribution for problems proposed to the children.

"There are four dogs in each house. Each dog will get a biscuit like the one drawn on the board. Draw the number of biscuits we need to have so that each dog gets a biscuit."
(NUNES et al., 2005, p. 88)

This problem was presented through drawings and oral instructions, and there was a gap between the percentage of correct answers when presented with materials that allowed the direct application of the action scheme and the percentage of correct answers when applied with pencil and paper.

As for the multiplication and division problems with the same structure:

> "Problem 1: "Marcio has invited three friends to his birthday party. He wants to give each friend 5 marbles. How many marbles does he need to buy?
> Problem 2: Marcio has 15 marbles. He's going to distribute them equally among his friends. How many marbles will each of them get?"
> (NUNES et al., 2005, p. 89)

It was observed that the children used the action schema of distributing to solve these problems. Subsequently, the children were given other problems like these and it was concluded that even first graders who had not been taught the mathematical concepts of multiplication and division used action schemes and solved the problems correctly. It also became clear that it is possible to improve the development of multiplicative reasoning if the concepts of multiplication and division based on action schemes are proposed, instead of using the addition and subtraction of equal parts as the key point.

Various studies have shown that the beginning of understanding the concept of division occurs long before formal teaching, so there is a need for teachers to know the action schemes that children use in the written solution of situations involving division, before formalising this concept. Vergnaud (1990) states that it is through situations and problems to be solved that a concept acquires meaning for the child.

In this research, we are going to approach the concept of division from the conceptual fields of multiplicative structures with problems of division, exact, inexact, partition and quotas, for children in the first grades of primary school.

CHAPTER 5

LITERATURE REVIEW

There are a number of studies aimed at verifying the issues surrounding the mathematical concept of division, and below we will highlight the studies with perspectives common to this research.

Research	Subjects	Objectives	Methodology	Conclusions
The use of concrete materials in solving division problems was discussed. Selva (1998)	One hundred and eight children, in literacy, first grade and second grade, from a public school in Recife.	• Evidence of children's understanding of the definition of division remainder • Analyses the influence that each type of representation (concrete, written and mental) can have on understanding the concept of division.	After the survey, the children were organised into three groups. The first group was given tokens of the same size and colour, the second group was given paper and pencils and the third group was not given any materials. Each child was asked to solve eight division problems, four partition problems and four quota problems.	The children performed better on the partitioning problems, but had difficulties to work with the rest. Regarding the use of the material, it was noted that it is interesting insofar as the child develops more advanced strategies, but it is no longer interesting for younger children who already have more flexibility in using paper and pencil strategies or mental calculation.
Multiplicative structures and raising awareness: sharing to share (Moro, 2005)	Six first and second grade students, aged between seven and eight, from a public school in the Curitiba region.	• Describe the children's conceptions of division by partition. • Identify levels of awareness of the concept of division.	They were grouped into triads and offered tasks with a division by partition problem situation, which were proposed orally by the researcher for the participants to solve together, according to their strategies. The materials used were: plastic tokens, a box with a divider divided into two halves, two dolls, sheets of cardboard and marker pens.	0 concept of division is centred on pre-additive conceptions, on dividing into smaller and then larger quantities and on distributing. The progress observed in each child's understanding of division by partitioning is linked to the "realisation" of schemes and relationships pertinent to the concept.

Table 1: Research on the same theme

Research s	Subjects	Objectives	Methodology	Conclusions
The relationship between division problems and children's conceptions of division (Lautert, Spinillo, 2002)	Eighty children aged between five and nine from public schools in Recife.	- To investigate children's mathematical knowledge of division from two perspectives: performance in division problems and conceptions of division.	The problems at the first stage were presented orally, being exact division, partitioning and quota problems. The children were given paper and pencil and were instructed to solve them in any way they wished. In the second phase, interviews of a clinical nature were carried out, in which the	The children who had received school instruction on division found it easy to solve the partition and share problems correctly, while the children who had not received instruction found it difficult. The children's definitions of what it means to divide revealed that they had no mathematical meaning, with mathematical meaning not associated with division and conceptions with a mathematical meaning directly

			question was asked: "What does it mean to divide?" so that the child could make mathematical sense of it.	associated with division.
Awareness-raising analysed from the concept of division: a case study. (Ferreira J Lautert, 2003)	A male child aged six years and four months, studying literacy at a public school in Recife.	- to discuss awareness through a case study of children solving division problems.	The child was asked to solve a division problem in two different situations: a graphic situation in which the child was given paper and pencil, and a concrete situation in which the child was given cards and objects similar to those described in the problem. The examiner read out the problem and instructed the child to represent it in any way they wanted.	They showed varying degrees of awareness of division, and did not reach conceptualisation, i.e. the idea that division refers to the idea of totality and interdependence between its terms. The children resorted to their already constructed addition schemes to solve the division problem.

Table 2: Research on the same theme

Research	Subjects	Objectives	Methodology	Conclusions
The oral resolution of division tasks by children (Corrêa , 2004).	Eighty-three children aged between six and nine attended a public school in Oxford (UK).	- to analyse the performance of children with different levels of schooling in arithmetic in the oral solution of division problems by partition and by quotas and to describe these strategies.	The children were presented with a situation in which a certain number of blocks represented the food that had to be shared between a certain number of bears, and the children had to say how many blocks each bear would receive individually. Four sizes of dividend (4, 8, 12 and 24) and two divisors (2 and 4) were used in this study.	The children's performance varied according to age and schooling. Double counting procedures and the use of multiplicative facts appeared more frequently in division by quota situations, while procedures based on repeated additions and partitioning of quantities appeared in division by partition situations. A small percentage also used procedures based on repeated subtraction.
Maths notes for children equalising and distributing large areas at the origin of multiplicative structures (Moro, 2004).	Twelve students, aged between six and ten, from two public schools in different municipalities, located on the urban outskirts of two large cities.	• Describe the nature and transformations of children's notations relating to the tasks of equalising portions and dividing up quantities, aimed at developing additive and multiplicative relationships; • Verify the meaning of the notations produced when examining additive relationships and multiplicative relationships.	The triads were drawn by lot according to the optimum lag criterion. The tasks (partition and quotient) were proposed orally by the researcher, with problem situations. The material used consisted of 18 plastic tokens (the same colour), a box with a divider dividing it into two halves, two dolls, sheets of cardboard and marker pens.	The children who had mastered division to a more advanced degree were still unaware of the arithmetic signs for division and the canonical ways of expressing division. They were not aware of the additive transformation relationship from an initial quantity to a final one. The construction of additive-subtractive relationships identified in the notations points to the importance and complexity of constructing schemes for equalising and unequalising portions. There was strong evidence of schemes for dividing quantities. The actions of

				equalising and unequalising, related to dividing up numerical quantities, were central to these studies.

Chart 3: Research on the same theme

In the literature, all the studies described involve division problems by partition, or by quotas, or both. Correa (2004) draws attention to the initial construction of the concept of division by the child, how important it is for the child to define these two classes of problems: division by partition and division by quotas. In Moro (2005) we realise how early division is among children and Selva (1998) describes that these problems are treated equally by pupils and that partition problems are considered easier than division by quota.

The literature generally shows that partitioning problems are easier for children than quotas. One of the explanations for this is the child's initial notion of division, which stems from their experience of dividing a whole into equal parts until there is no more to divide. Notions of division stem from the idea of distributing, as shown by (Selva, 1998; Moro, 2004 and 2005; Lautert and Spinillo, 2002 and Ferreira, Lautert, 2003).

Of the studies observed, the one with the most similarities to this study is that of Lautert and Spinillo (2002), because the methodology used is similar to the one used in this research: the problems will be presented orally, with inexact and exact division, partitioning and quota problems. The children will be given paper and pencils and will be instructed to solve the problems as they wish.

CHAPTER 6

METHODOLOGY

In order to achieve the proposed objectives, a mathematical knowledge form with exact and inexact division, partition and quotient problems was used to gather information on the object being investigated. The qualitative analysis of the records used by the subjects in solving the division problems made it possible to learn about the variety of strategies used and, in some cases, to infer the concepts and action schemes used by the subjects.

6.1 Subjects

This exploratory study was initially carried out with 38 primary school students aged between six and seven in two morning classrooms at a municipal school in Navegantes, Santa Catarina.

We realised that most of the previous studies investigating this topic were carried out with subjects with little schooling, generally from schools located in poor communities. This choice allows us to observe the use of diversified strategies, since these subjects have not been so influenced by school teaching, which tends to homogenise problem-solving procedures. In view of this fact, it was also decided to choose schools with the same socio-economic characteristics.

6.2 Instruments

To collect the data, a form was drawn up with problems involving the concept of division, exact and inexact division, partitioning and quotients. This form was designed to work with three variables, thus creating a research design that would allow the effect of each of the variables to be verified. In other words, the problems were differentiated in terms of unit (continuous vs. discrete), type of problem (partition vs. quotient) and whether or not there was a remainder, as shown in the table.

Problems	Quantity	Classification (type)	Resto
PO1	Continuous	Partition	No
PO2	Discreet	Partition	Yes
PO3	Discreet	Quota	Yes
PO4	Discreet	Partition	No

Table 4: Classification of problems

Comparing the problems makes it possible to check the effect of the variables, as follows:

- Type of quantity: continuous x discrete (PO1 x PO4) - unit effects

- Type of problem: partition x quotas (PO2 x PO3) - type effects
- Whether or not there is leftovers (PO2 x PO4) - effects of leftovers

The form was drawn up with the following division problems:

1. POl - Division by Partition (exact): "A friend of mine came home **starving and made two sandwiches out of bread, ham and cheese. Just as he was about to take the first bite, three of his friends arrived. There was no more bread, ham and cheese. What would you do if you were in** his **place**?" This problem is taken from the book: Problems? What Problems? by Mercedes Carvalho (2005), where the author deals mainly with teachers' complaints that students don't know how to interpret problems. This problem involves the concept of fractions: two integers divided by four. The problem statement does not mention the quantities directly.

In this case, the dividend is the two sandwiches, the divisor is the protagonist and his three friends, i.e. four, and the expected quotient can be either one piece of bread (1/2) or two pieces of bread (2/4). For this problem, the remainder is zero, or there is no remainder. The expected action scheme for solving this problem (according to NUNES et al., 2005) should be that of equitable distribution, since the relationship between the whole and the parts is unknown.

2. P02 - Division by Partition (inaccurate): **Peter had bought 16 trolleys and had 5 boxes. He wanted to put the same number of trolleys in all the boxes. How many trolleys did he have to put in each box?** This problem is similar to the problems that Selva (1998) applied to literacy children when the aim was to investigate the resolution of division problems with a non-zero remainder, i.e. inexact division problems.

In the statement of this problem we have the dividend: the sixteen trolleys and the divisor: the five boxes, and what is expected is the quotient, three trolleys in each box with a remainder equal to one, i.e. there would be one trolley left over.

To make it easier to check which action scheme the child used in their written solution, we organised the problems (PO2, PO3 and PO4) into tables, initially assigning different values to the variables to identify the fixed relationship in the problem, as in the examples by Nunes et al. (2005).

PO2 - Partition	**Problem data**		
Pedro had bought 16 trolleys and had 5 boxes. He wanted to put the same number of trolleys in all the boxes.	No. of boxes	No. of trolleys per box	No. of trolleys

How many trolleys did he have to put in each box?	1	3	3
	2		6
	3		9
	4		12
	5		15 (+1)
	:		:

Table 5: Denotation of the fixed PO2 ratio.

From this table, we have no doubt that the fixed ratio in this problem is the number of trolleys per box, in this case three, which translates into the number of trolleys for each box. As this fixed ratio is not described in the problem, it is not known, so the expected course of action to get the correct answer to this problem must be that of equitable distribution.

3. PO3 - Division by quotient (inexact): **Marta had 19 sweets and wanted to put 6 sweets on each tray. How many trays will she need?**
   According to Vergnaud (1994), there are two basic types of division problems: isomorphism problems (called partition problems) and division by quota problems. In this situation, we have a division by quota problem, like the example discussed in the theoretical framework. However, in this case, it is also an inexact division.

In the problem statement we have the dividend: nineteen sweets and the quotient: six sweets in each tray, and the expected divisor is three trays and the remainder is one, one sweet.

PO3 - Quota	**Problem data**		
Marta had 19 sweets and wanted to put 6 sweets on each tray. How many trays will be necessary?	No. of trays	No. of sweets per tray	No. of sweets
	1	6	6
	2		12
	3		18 (+1)

Table 6: Denotation of the fixed ratio of PO3.

In this problem, we realise that the fixed relationship is the number of sweets per tray, which in this situation is six, and this number is expressed in the statement. Therefore, the action scheme expected to solve it correctly is one-to-many correspondence.

4. Division by Partition (exact): **Júlia has 6 chocolates and wants to divide them between 3 friends. How many sweets will each friend get?** This problem, according to Vergnaud, is a partition problem, an exact division problem.

In the problem statement we have the dividend, the six sweets, and the divisor is the three friends, and the expected is the quotient, two sweets for each friend, and the remainder is zero, there's nothing left.

PO4 - Partition	Problem data		
Julia has 6 chocolates and wants to share them with 3 friends. How many sweets will each friend get?	No. of friends	No. of chocolates per friend	No. of chocolates
	1		2
	2	2	4
	3		6

Table 7: Denotation of the fixed ratio of PO4.

The fixed ratio in this problem is the number of sweets per friend, in this case two. And this number is not known in the problem statement either. So as with PO2,

the expected course of action (also according to NUNES et al., 2005), which will indicate the correct answer to this problem, is that of equitable distribution.

6.3 Data collection procedures

After the project was approved by the ethics committee, the school was presented with an authorisation form to carry out the research (shown in appendix 1) and parents were also sent an authorisation form for their children to take part in the research (shown in appendix 3).

To preserve the identity of the subjects, the forms were coded.

The data was collected by the researcher herself in the school environment. The forms were applied to each class, one on each day. The forms were applied and given to all the students present in the classroom at the time of application (38 subjects). The researcher read out each problem and waited until everyone had finished recording their solutions before presenting the next problem. The researcher and the class teachers, who were present at the time of the application, did not interfere in the subjects' solutions. The application took approximately one and a half hours in each class,

The subjects were allowed to use paper and pencil and were told that they could solve the problems in any way they wished.

The qualitative analysis of the written solution strategies employed by the subjects in solving the division problems indicated whether the schemes of action, equitable distribution and one-to-many correspondence were used by the students in the first grades.

CHAPTER 7

ANALYSING AND DISCUSSING THE RESULTS

7.1 Organising the data

In the written solutions, most of the pictorial and symbolic registers were used to solve the proposed division problems. Some children used numerals in their solutions simply to represent the division factors, but did not use numerals to represent the operation. Thus, the algorithm for solving division was not found in any of the solutions, perhaps due to the age group of the children. Others wrote down the answer to the problem in their mother tongue and others used different forms of representation, for example pictorial or symbolic representation and numerals. We note that the way in which the children presented their records when solving the problem is not related to the solution strategies adopted, which is why the type of representation was not taken into account when analysing the data.

In order to characterise the written solutions used by primary school students when solving division problems, we first tried to quantify the information in a table. Since it was not possible to visualise the strategies in this way, we decided to draw up a graphic diagram (in the format of an organogram) to make the path used clearer, which would make it easier to identify the action schemes. The graphic diagram was organised according to the strategies that were identified in the written solutions after they had been analysed.

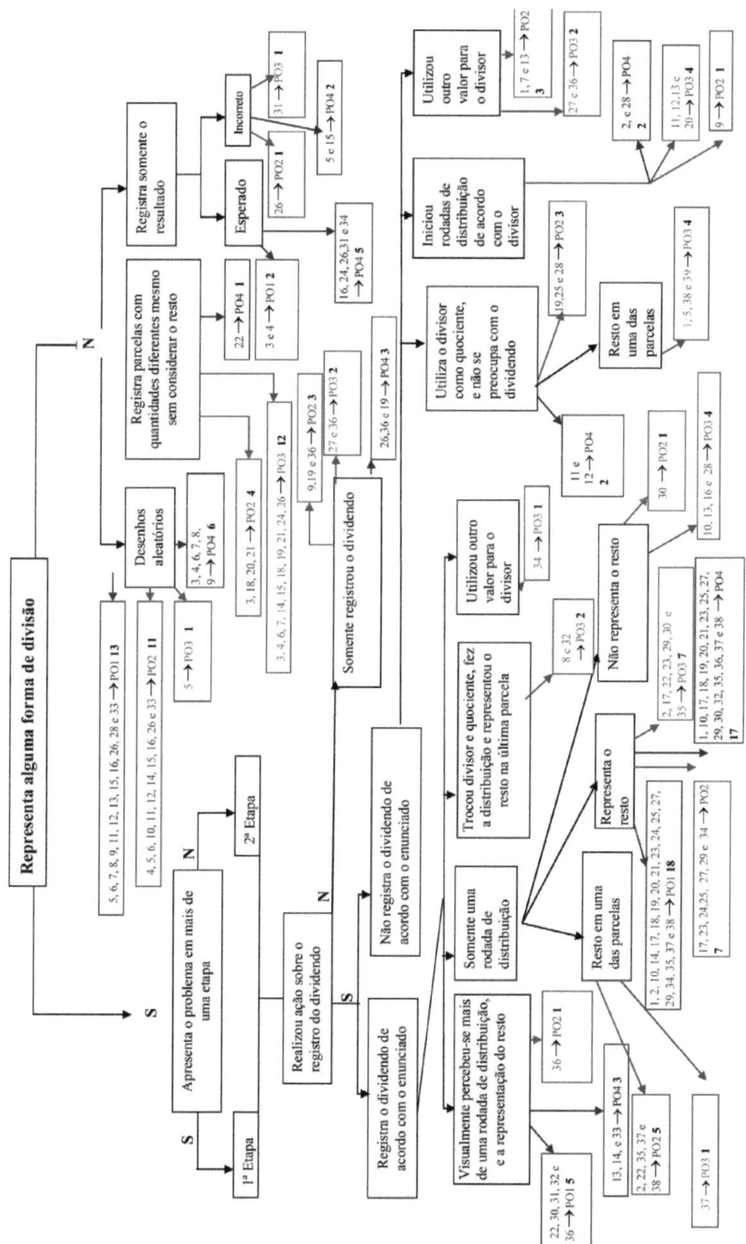

Illustration 1: Graphic diagram of problem-solving strategies

In the graphic diagram, the coloured end rectangles are to designate the problem of the data collection form, the coloured numbers identify the children and the black number at is the number of children who followed this "path".

We put the following observations as key points:

1. It represents some form of division:

    - if the child didn't represent some form of division, we found the following situations in the records: random drawings; parcels with different quantities, even without considering the remainder; and records of the result only.

    - whether they presented some form of division, whether they presented the solution to the problem in more than one step, whether they acted on the dividend, in which case they only recorded the dividend, whether they recorded the dividend according to the statement or not.

We noticed that the children who didn't record the dividend according to the statement used the following procedures to continue solving the problems: they used another value for the divisor; they started distribution rounds according to the divisor; they used the divisor as the quotient and didn't worry about the dividend.

When they represented the dividend according to the statement, we saw the following situations: they used a different value for the divisor; they swapped divisor and quotient; they distributed and represented the remainder in the last instalment; they only did one round of distribution; and visually we saw more than one round of distribution and the representation of the remainder. Among the situations that led to a form of division, they were also organised in terms of the representation of the remainder.

7.2 Analysing the procedures

In order to clarify the children's location on the map, we will report on the organisation of the sequences and comment on them. We began by identifying whether the child represented some form of partition in their written solution.

In 61 of the solutions (38.13%), out of the 160 observed, because some children represented the solution in two stages, no form of division was identified, and by problem we have the following totals according to the procedures:

Procedures	Problems				Total
	PO1	PO2	PO3	PO4	

		F	FR(%)	F	FR(%)	F	FR(%)	F	FR(%)	F	FR(%)
Records random drawings		13	**8,12**	11	**6,80**	1	**0,63**	6	**3,75**	31	**19,38**
Record parcels with different quantities, even without consider o rest.				4	**2,50**	12	**7,50**	1	**0,63**	19	**10,63**
Record only the result of	Waiting for	2	**1,25**	-	-	-		5	**3,13**	7	**4,38**
	Incorrect	-	-	1	**0,63**	1	**0,63**	2	**1,25**	4	**2,50**
Total		15	**9,47**	16	**9,93**	14	**8,76**	14	**8,76**	61	**38,13**

Chart 8: Quantification of the procedures that did not represent signs of division.

In those who didn't, we realised that:

- They simply recorded random drawings without the idea of partitioning, for the first problem (POl) of the form approximately *9%* of the children's solutions, for the second (PO2) problem 7%, for the third (PO3) approximately 1%, which corresponds to only one child and for the fourth (PO4) problem 4% of the children, totalling 19.38% of the solutions. Example of this procedure:

Illustration 2: Written solution of one that simply represented random drawings, for the first problem of the form.

- They recorded parcels with different quantities, even without considering the remainder, i.e. the parcels had different numbers of elements. We identified 2.5 per cent of the children's solutions to the PO2 problem, 7.5 per cent to the PO3 problem and 0.63 per cent to the PO3 problem.

for the fourth problem, only one child. Here's an example of this procedure:

Illustration 3: Written solution for the third problem on the form by a person who represented parcels with different quantities, even without considering the remainder.

- They only recorded the result, and in these solutions it wasn't possible to check the solving strategy or the form of partitioning. In this case, we looked at whether they recorded the expected result, for problem PO1 1.25% of the solutions and for problem PO4, 3.13%. We also found that the children who only recorded the incorrect result for PO2 and PO3 corresponded to 0.63% for each problem, and for PO4, 1.25%. Let's see an example of this procedure:

Illustration 4: Written solution of a child who represented only the result for the second problem of the form.

In this stage of the map, 15 solutions from PO1, 16 solutions from PO2, 14 from PO3 and PO4 were discarded from the 160 solutions that were presented in total, because they didn't show division records. Solutions that only described the expected result of the problem were discarded because the child must have done some reasoning to reach that conclusion and we were unable to classify the strategies due to the lack of information.

We noticed that in the inexact partitioning problem, the children had more difficulty representing some form of division, which didn't happen so often in the exact partitioning PO4. This is because the situation described in the problem statement is directly related to the students' experience, because it is an exact partition problem and also because it presents small quantities (six sweets) to be distributed among three friends.

2. In those that represented some form of division, we checked whether the problem was presented in more than one stage, and we observed in a few solutions (eight solutions) that the children presented some problems (PO2, PO3 and PO4) in two stages of resolution, not being the same children. Only one child acted on both registers of the dividend.

In 8 solutions, equivalent to 5% of the children's solutions, we noticed that they only recorded the dividend, in 21 solutions, corresponding to 13% of the solutions, we noticed that they recorded the dividend with a different value to those described in the problem statement and in 72 solutions, corresponding to 45% of the solutions, we verified that the dividend was recorded according to the statement.

Of those children who performed some action on the dividend they represented, we have:

- When they didn't, we noticed that they only represented the dividend, but didn't use this record for other actions, i.e. to distribute or share.

At this stage of the map, we discarded three more solutions from PO2, two from PO3 and three from PO4, as no idea of division was identified in the written solutions. We'll see examples of these procedures:

Illustration 5: Written solution of one who represented only the dividend for the fourth problem of the form.

- When they acted on the record of the dividend, we checked whether the dividend represented was the one expected for the problem or incorrect.

1. In the event that the dividend has been represented incorrectly according to the problem statement, we find the following procedures for problems PO2, PO3 and P04.

Procedures	**Problems**						**Total**	
	PO2		**PO3**		**PO4**			
	F	**FR(%)**	F	**FR(%)**	F	**FR(%)**	F	**FR(%)**
Used other value for the divisor.	3	**1,88**	2	**1,25**	-	-	5	**3,13**
Started distribution rounds according to the divisor.	1	**0,63**	4	**2,50**	2	**1,25**	7	**4,38**
Uses the divisor as the quotient and doesn't worry about the dividend.	3	**1,88**	4	**2,50**	2	**1,25**	9	**5,63**
Total	7	**4,39**	10	**6,25**	4	**2,50**	21	**13,14**

Chart 9: Quantification of the procedures that represented the incorrect dividend, according to the problem statement.

He used another value for the divisor, in which case, as the dividend and divisor were represented with different values to those described in the problem statement, we can't say that these written solutions don't have strategies that show forms of distributions, but we can't characterise the schemes present in these solutions, as the results are far from those expected for these problems. In this stage of the graphical scheme, 1.88% of the solutions for PO2 and 1.25% for PO3 are found. Since these problems are inexact, PO2 is a partition problem and PO3 is a quota problem, this justifies the use of this procedure. We'll see an example of this procedure.

Illustration 6: Written solution of one who represented the divisor differently from the value described in the statement for the third problem.

=> Started rounds of distribution according to the divisor, in these solutions we realise that the child didn't bother to exhaust the dividend described in the statement, thus doing a few rounds of distribution. Similarly, we don't characterise these schemes because the solutions are not the ones expected for the problems, and the result is far from what was expected. At this stage of the map, 0.63 per cent of the solutions were for PO2, 2.50 per cent for PO3 and 1.25 per cent for PO4. The use of this procedure occurred more frequently in the quota problem (PO3). Perhaps this is the reason for the incorrect solution and the use of this procedure, because according to the literature (SELVA, 1998) quota problems are considered more difficult by children....

Illustration 7: Written solution by a child who represented the divisor according to the statement but didn't bother to exhaust the dividend described in the fourth problem.

=> Uses the divisor as the quotient and doesn't worry about the dividend, in this situation, according to Nunes et al. (2005), the child makes a one-to-many correspondence, because they use the divisor as the quotient and make the distributions without worrying about all the parts, i.e. the dividend. This occurred in 1.88% of the solutions for PO2, 2.50% for PO3 and 1.25% for PO4. Let's look at an example of this procedure

Illustration 8: Written solution of a child who used the divisor as the quotient and didn't worry about the dividend when solving the fourth problem.

2. In the event that the dividend represented is the one expected to solve the problem according to the statement, we quantify the following procedures:

Procedures	Problems								Total	
	PO1		PO2		PO3		PO4			
	F	FR(%)	F	FR(%)	F	FR(%)	F	FR(%)	F	FR(%)
Used another value for the divisor	-	-	-	-	1	0,63	-	-	1	0,63
Swapped divisor and quotient, did the distribution and represented the remainder in the last instalment					2	1,25			2	1,25
There was only one round of of distribution	18	11,25	13	8,12	12	7,50	17	10,63	60	37,5
More than one round of distribution was visually noticeable	5	3,13	1	0,63	-		3	1,88	9	5,64
Total	23	14,38	14	8,76	15	9,38	20	12,43	72	44,57

Chart 10: Quantification of the procedures that represented the dividend according to the problem statement.

Thus we have

=>He used another value for the divisor, which was greater than the one described in the problem statement. Only one child in their solution (0.63%) to PO3 used five as the divisor, and put four in the first four instalments and three in the last instalment, totalling nineteen, which is the value described in the statement for the dividend. We haven't identified why she didn't use the value of six described in the statement for the quotient, but the procedure is justified because of the quota problem.

In this case, the result is far from what was expected, so we haven't categorised the scheme used in this resolution either.

Illustration 9: Written solution of a child who used the divisor with a value greater than the one described in the statement of the third problem.

=>In this situation, two children in their solutions to P03, which corresponds to 1.25% of all solutions, used the value described in the statement, six sweets for each tray, as the divisor, and placed three sweets in the first five trays and four sweets in the last parcel. We realised that the scheme used for this solution was that of equitable distribution, which made the solution to this problem incorrect according to the Nunes et al. literature (2005). The problem had a fixed ratio of six sweets per tray, which characterises the use of the one-to-many scheme.

Illustration 10: Written solution of a child who swapped the divisor and quotient and did the distribution in the solution to the third problem.

=> Only one round of distribution: visually, the similarity of the drawings shows that the children did only one round of distribution. What may justify this strategy is the wording of the problem, which translates a situation experienced by the students. In this way, they arrive at the value of the quotient even when it is unknown in the problem statement. In other words, when the fixed ratio is not present in the problem statement, these division problems are classified as partition problems. In this procedure, we characterised the partition problems as solved using the equitable distribution scheme and the quota problem as solved using the one-to-many scheme, as described in the literature (NUNES, et al. 2005), and as expected for solving these problems. This procedure led to the classification of the remainder as: remainder in one of the parcels; represents the remainder; and here PO1 and PO4 were also classified as having a remainder equal to zero, and the situation in which the child does not represent the remainder. Examples of this procedure:

Illustration 11: Written solution of one who visually represented only one round of distribution and placed the rest in one of the plots for the second problem.

Illustration 12: Written solution of one who visually represented only one round of distribution and represented the rest correctly for the second problem.

Illustration 13: Written solution of one who visually represented only one round of distribution and didn't represent the rest for the third problem.

In this procedure we obtained 11.25% of the solutions presented for PO1, 8.12% for PO2, 7.50% for PO3 and 10.63% for PO4, all of which were correct for the problems in question. In this procedure, we characterised the action schemes as equitable distribution for partition problems and one-to-many correspondence for quota problems, according to (NUNES et al., 2005).

As for the expected answer to the problem, we only considered correct the situations in which the child represented the remainder, which included the solutions to PO1 and PO4; that the child did not represent the remainder and the solutions in which the remainder was represented in one of the parcels were considered incorrect for the problem.

Procedures	Problems								Total	
	PO1		PO2		PO3		PO4			
	F	FR(%)	F	FR(%)	F	FR(%)	F	FR(%)	F	FR(%)
Remainder in one of the instalments	-	-	5	**3,13**	1	**0,63**	-	-	6	**3,75**
Represent the rest	18	**11,25**	7	**4,38**	7	**4,38**	17	**10,63**	31	**30,64**
It doesn't represent the rest	-	-	1	**0,63**	4	**2,50**	-	-	5	**3,13**

Chart 11: Quantification of the procedures relating to the representation of the remainder for the problems.

=> Visually, more than one round of distribution and the correct representation of the remainder were perceived. In this situation, we can see from the blots and blackouts that the child made more than one round of distributions, characterising at first the one-to-one correspondence, characterising the equitable distribution scheme, Nunes et al, (2005). And this procedure was only present in the partitioning problems, with 3.13% of the solutions presented being for PO1, 0.63% for PO2 and 1.88% for PO4. It is not present in the solutions to the quota problem, as this procedure would not be appropriate for solving this problem because the fixed ratio is known. Let's look at an example of this procedure:

Illustration 14: A child's written solution to the fourth problem.

In this way, we characterised the expected schemes in the written solutions in which division ideas were presented to children in the first grade of primary school, where we found that these expected action schemes resulted from procedures in which only one round of distribution was evident or more than one round of distribution was visually perceived.

Procedures	Probemus								Total	
	PO1		PO2		PO3		PO4			
	F	FR(%)	F	FR(%)	F	FR(%)	F	FR(%)	F	FR(%)
There was only one round of distribution **(the schemes observed were for distribution fair e one-to-many correspondence)**	18	**11,25**	13	**8,12**	12	**7,50**	17	**10,63**	60	**37,5**
Visually if realised more than one	5	**3,13**	1	**0,63**			3	**1,88**	9	**5,64**

round of distribution (the schemes observed were distribution equitable)										
Total	23	14,38	14	8,75	12	7,50	20	12,51	79	43,14

Chart 12: Quantification of the procedures in relation to the schemes expected to solve the problems.

So 14.38% of the solutions presented were correct according to the expected scheme for PO1, 8.75% for PO2, 7.50% for PO3 and 12.51% for PO4, totalling 43.14% of the expected solutions according to the action scheme.

As for the expected representation of the division terms, we can see in the graphic diagram that for the dividend to be represented correctly, the child would need to reach the procedures for correctly recording the dividend and also those who represented only the dividend, which corresponds to 80 solutions, exactly 50 per cent.

The divisor, on the other hand, was represented according to the problem statement only for the children who arrived at the following procedures: visually, more than one round of distribution, and the representation of the remainder; only one round of distribution and initiated rounds of distribution according to the divisor, totalling 76 solutions, which corresponds to 47.5% of the solutions.

The quotient was represented in accordance with what was expected for the wording of the sixty problems only for the children who arrived at the procedures: visually more than one round of distribution was perceived and the representation of the remainder and only one round of distribution, totalling 69 solutions corresponding to 43%.

7.3 Effect of variables

Analysing PO1 and PO4, we see the effect of the type of quantity described in the problem statement. In PO4 we have discrete quantities, and in PO1 we have continuous quantities, but the success rate was higher in PO1 (14.3 per cent) as shown in the table above. (It is likely that the situation described in the PO1 statement is one that the child experiences frequently when sharing a snack with friends or siblings. As Vergnaud (1990) reminds us, the familiarity of situations allows children to draw on the schemes they have already developed to solve similar situations and apply them to solving the new problem.

We noticed this in PO1, whose wording is different from the problems that are emphasised in textbooks, and which refers to a situation where the child imagines themselves in the context of the wording. This was evidenced by the fact that the children represented this problem predominantly using the pictorial register and this problem was the one in which the children showed the best rate of achievement.

To check the effect caused by the type of division problem, we compared the solutions to the second (PO2) and third (PO3) problems. The inexact partition problem (PO2) showed a better level of learning (8.75% of the answers) than the inexact partition problem (7.5% of the answers), as stated in the literature by Selva (1998). Even so, we concluded that the children's performance in solving exact partition division problems was the best.

Type of problem		**Answers considered correct according to the statement**	
		F	FR(%)
Partition	Exact (POl and PO4)	43	26,88
	Inexact(PO2)	9	5,63
Dues	Inexact(PO3)	11	6,88

Chart 13: Quantification of results in relation to the type of division problem.

Analysing the effect caused by the remainder, we observed that children in this age group have difficulty representing the remainder when it is different from zero, as presented in the literature by (LAUTERT, SPINILLO, 2002). When we compared PO2 with PO3, and considered the effects of the remainder, we realised the best rate of correct answers for PO3 (6.88%), which is a problem of inexact quota.

CHAPTER 8

FINAL CONSIDERATIONS

The qualitative analysis of the subjects' written solutions to the division problems resulted in a map that lists the stages that led to the strategies that resulted in the action schemes used by the children to solve the division problems. And also in the coordination of the factors dividend, divisor and quotient involved in the situation arising from the problem statement. After a few attempts to organise the data in a table, it was realised that this way it wouldn't be possible to visualise all the information, so we opted for a map.

The research was carried out with children attending the first grade of primary school at a municipal school in Navegantes, with the aim of revealing an understanding of division in their written solutions. In order to do this, it was necessary to characterise the written solutions used by primary school pupils when solving division problems, and therefore the following specific objectives were set: 1. to check whether the written solutions contain the terms of the division (dividend, divisor) and the result (quotient), and whether they are consistent with the information in the problem statement; 2. to identify whether the action schemes: equal distribution, equal distribution, equal distribution) are present in the written solutions. To identify whether the action schemes: equitable distribution and one-to-many correspondence according to Nunes, Campos, Magina and Bryant, (2005), are present in the written solutions of students in the first grade of primary school; and 3. To check whether the written solutions reveal the existence of coordination between the schemes.

Looking at the research data, we realise that the division revealed in the children's solutions is directly related to the correspondence and distribution schemes used by them, as they have no concept of the division algorithm due to their age group. And these schemes are influenced by the situations described in the division problem statements.

As for the representation of the division factors (dividend, divisor and quotient), we observed that the majority of the children represented the dividend as expected, while the divisor and quotient were close (50%, 47.5% and

43%)according to the expected value, it was observed that the children who represented all the factors expected for the problem managed to obtain the expected result.

It can be concluded that the understanding presented in the expected solution of the problems at is directly linked to the classification of the problems, but rather to the situation described in the statement. In general, the children's performance in solving exact partition problems was the best, while the inexact partition problem showed a better level of learning than the inexact partition problem, when the remainder was taken into account. These factors are extremely interlinked with the problem situation: when the problem is written in a way that is appropriate to the child's reality, where the child refers to the story described in the problem statement, they are able to solve it more easily. We realised this in the first problem, which has a different wording to the problems in textbooks, and which refers to a situation where the child is transferred to the context of the wording, and this problem was the one that the children showed the best rate of achievement.

The researcher hoped and still hopes that, from this research, teachers will get to know the understanding revealed in children's solutions to division problems, reflect on them and, if possible, promote teaching that leads students to develop and know the skills needed to understand this concept and this subject.

CHAPTER 9

REFERENCES

BICUDO, Maria Aparecida Viggiani. **Research in maths education: Conceptions and Perspectives.** São Paulo: Cortez, 1999.

BRASIL. **Parâmetros Curriculares Nacionais (I[a] a 4a série): matemática/Secretaria de Educação.** Fundamental Education. Brasília: MEC/ SEF, 1997.

BRUN, Jean. **Didactics of Maths.** Lisbon: Piaget Institute, 1996.

CARVALHO, Mercedes. **Problems? But what problems?: Strategies for solving mathematical problems in the classroom.** Rio de Janeiro: Vozes, 2005.

CORRÊA, Jane. **Children's oral resolution of division tasks.** A case study. Psicologia: Reflexão e Crítica, Natal, v. 9, n. 1, 2004. Available at: http://www.scielo.br/scielo.php.

CORRÊA, Jane & MEIRELES, Elisabet de Souza. **Children's intuitive understanding of the partitive division of continuous quantities.** Estudos de Psicologia, Natal, v. 5, n. 1, 2000. Available at: http://www.scielo.br/scielo.php.

FALCÃO, Jorge Tarcísio da Rocha. **Psychology of Maths Education.** Belo Horizonte: Autêntica, 2003.

FERREIRA, Sandra Patrícia Ataíde & LAUTERT, Síntria Labres. **Consciousness-raising analysed from the concept of division.** Psicologia: Reflexão e Crítica, Porto Alegre, v. 16, n. 3, 2003. Available at: http://www.scielo.br/scielo.php.

INEP. **Releases PISA/2006** (Programme for International Student Assessment) **Results** available at: www.inep.gov.br.

LAUTERT, Síntria Labres; SPINILLO, Alina Galvão. **The relationship between performance on division problems and children's conceptions of division.** Psicologia: Teoria e Pesquisa, Brasília, v. 18, n. 3,2002. Available at: http://www.scielo.br/scielo.php.

MORO, Maria Lucia Faria. **Notations in children's maths. Equating and dividing quantities at the origin of multiplicative structures.** Psicologia: Reflexão e Crítica, Porto Alegre, v. 17, n 2,2004. Available at: http://www.scielo.br/scielo.php.

MORO, Maria Lucia Faria. **Multiplicative structures and awareness: dividing to divide.** Psicologia: Teoria e pesquisa, Porto Alegre, v. 21, n 2,2005. Available at: http://www.scielo.br/scielo.php.

NUNES, Terezinha. & BRYANT, Peter. **Children doing maths.** Porto Alegre: Artes Médicas, 1997.

NUNES, Terezinha; CAMPOS, Tânia Maria Mendonça; MAGINA, Sandra & BRYANT, Peter. **Maths education: numbers and numerical operations.** São Paulo: Cortez, 2005.

PESTANA, Maria Inês. **Prova Brasil results and the challenges for municipal leaders: the obvious challenge, improving the quality of Brazilian education.** Available at: http://www.undime.org.br/htdocs/download.php

PIAGET, Jean. & SZEMINSKA, A . **A génese do número na criança.** (C. M. Oiticica, Trad.) Rio de Janeiro: Zahar, 1971.

PIAGET, Jean. **The birth of intelligence in the child.** Rio de Janeiro: Zahar, 1970.

PIAGET, Jean, GRECO, Pierre. **Learning and knowledge.** Rio de Janeiro: Freitas Bastos, 1974.

PIAGET, Jean, INHELDER, B. **O desenvolvimento das quantidades físicas na criança.** Rio de Janeiro: Zahar, 1975.

PIAGET, Jean. **Six Studies in Psychology.** Rio de Janeiro: Forense Universitária, 1978.

PIAGET, Jean. **Biology and Knowledge.** Rio de Janeiro: Vozes, 1996.

SELVA, Ana Coelho Vieira. **Discussing the use of concrete materials in problem solving and division.** In A. Schliemann & D. Carraher (Orgs.), The comprehension of arithmetic concepts. Teaching and research. Campinas: Papirus, 1998.

TOLEDO, Marília, Toledo, Mauro. **The construction of maths.** São Paulo: FTD, 1997.

VERGNAUD. Gérard. **La théorie des champs conceptuels.** Recherches en Didactique des Mathématiques, 10(23): 133-170,1990.

VERGNAUD, Gérard. **The child, maths and reality: problems of teaching maths in primary school.** Mexico: Trillas, 1991.

VERGNAUD, Gérard. **Multiplicative conceptual field: what and why?** In Guershon, H. and Confrey, J. (1994). (Eds.) The development of multiplicative reasoning in the leaming of mathematics. Albany, N.Y.: State University of New York Press, pp. 41-59, 1994.

VERGNAUD, Gérard. The **web of conceptual fields in the construction of knowledge.** Revista do GEMPA, Porto Alegre, N° 4: 9-19,1996b.

VERGNAUD, Gérard. **Interview with Gerárd Vergnaud in Pátio.** Pátio Magazine, Porto Alegre, N° 5: 22-26, 1998.

WEBER, Demétrio. **Schools affiliated to FDG stand out in the Prova Brasil.** O Globo newspaper, 2008. Available at: http://www.fdg.org.br

WEINBERG, Mônica. **Education: It's going badly.** Article in Veja magazine, ed. 1884, 15 December 2004.

Printed by Books on Demand GmbH, Norderstedt / Germany